Separations Division 2019

Held at the 2019 AIChE Spring Meeting and
15th Global Congress on Process Safety

New Orleans, Louisiana, USA
31 March - 4 April 2019

ISBN: 978-1-5108-8692-6

Printed from e-media with permission by:

Curran Associates, Inc.
57 Morehouse Lane
Red Hook, NY 12571

Some format issues inherent in the e-media version may also appear in this print version.

Printed by Curran Associates, Inc. (2019)

For permission requests, please contact AIChE
at the address below.

AIChE
120 Wall Street, FL 23
New York, NY 10005-4020

Phone: (800) 242-4363
Fax: (203) 775-5177

www.aiche.org

Additional copies of this publication are available from:

Curran Associates, Inc.
57 Morehouse Lane
Red Hook, NY 12571 USA
Phone: 845-758-0400
Fax: 845-758-2633
Email: curran@proceedings.com
Web: www.proceedings.com

TABLE OF CONTENTS

550560 Industrial Distillation - Theory to Practice

This paper deals with a short history, and the basics of distillation as a Unit Operation. Distillation theory is converted into practice and how to optimally simulate and choose the appropriate number of stages for a tower.

550570 Dividing Wall Columns - a Primer

This paper presents an introduction to dividing wall columns and their applications in the chemical process industry

550899 Comparing Membrane & Distillation on an Equivalent Basis

Comparing Membrane & Distillation on an Equivalent Basis

Adrian Chavez Velasco, Rakesh Agrawal
Davidson School of Chemical Engineering, Purdue University, West Lafayette, IN

Among the separation technologies portfolio, distillation continues to be the preferable method for majority of the separations in the Chemical and Petrochemical plants. Just in the U.S. alone, it is estimated that about 90-95 % of all separations in chemical plants and refineries are performed by means of distillation columns [1]. Although distillation is widely used, it is frequently perceived to have a low energy efficiency, whereas in contrast, membrane separations are classified as a highly energy efficient processes which could potentially replace distillations on a wide scale [2], [3] [4].

However, this conclusion is not well supported since it has been mainly derived from comparisons made only with respect to conventional distillation, ignoring other operation modes that could be more energy efficient. Apparently, there is a widespread confusion in considering distillation as an absolute heat driven device, which sometimes requires a high amount of thermal energy input at the reboiler. The comparison between work driven membranes and heat driven distillations often leads to inaccurate conclusions. In reality, it is feasible to operate distillation solely with work input without requiring any external heat [5]. Our aim is to compare solely work driven distillations with membranes to avoid comparing two different forms of energy with each other.

The current research work centers on the development of a consistent procedure to compare the energy performance between membranes and distillation. This methodology comprises the use of rigorous simulation, and the optimization of both processes as an attempt to evaluate them at their best operating conditions. The proposed method was applied to the analysis of two important separation cases; p-xylene/o-xylene and propane/propylene. The results showed that the use of membranes with demonstrated permselectivity values does not always yield to the highest energy efficiency. Indeed, for the studied mixtures, the operation of distillation with work input consumes about an order of magnitude less energy than membranes for the case of p-x/o-xylene and ~27% less energy for the separation of propane/propylene under the conditions encountered at industrial scale (high recovery and purity). Membranes certainly offer some advantages over distillation in terms of energy efficiency, but for the analyzed cases, it only happens when the recovery is moderate or when the inlet is quite enriched in the most permeable component.

The extension of the developed methodology for the separation of other mixtures would enable researchers and practitioners to correctly identify under which conditions the use of either membranes or distillation would bring into the highest energy savings among them, thus contributing to design and operation of efficient separation processes.

Bibliography

[1] U.S. Department of Energy, "Materials Research for Separation Technologies: Energy and Emission Reduction Opportunities," 2005.
[2] D. S. Sholl and R. P. Lively, "Seven chemical separations to change the world," Nature, vol. 532, pp. 435-437, 2016.
[3] J. W. Koros, "Evolving Beyond the Thermal Age of Separation Processes: Membranes Can Lead the Way," AIChE Journal, vol. 50, no. 10, pp. 2326-2334, 2004.
[4] R. P. Lively and D. S. Sholl, "From water to organics in membrane separations," Nature Materials, vol. 16, pp. 276-279, 2017.
[5] A. A. Shenvi, D. M. Herron and R. Agrawal, "Energy Efficiency Limitations of the Conventional Heat Integrated Distillation Column (HIDiC) Configuration for Binary Distillation," Ind. Eng. Chem. Res., vol. 50, pp. 119-130, 2011.

550597 Column Inspections and Safety

Distillation column inspections are an essential step to a successful installation or revamp. Safety, mechanical design and knowledge of the process intended for the column are just a few of the topics that must be covered before entering the column. Setting up a checklist of items to be checked alone often requires significant attention to detail. This presentation will provide an introduction to preparation for a safe and successful column inspection.

550621 Distillation Tower Troubleshooting Tips for Beginners

Have you been called upon to troubleshoot a problem tower, but do not know where to begin?

This presentation is aimed at providing beginner tips and techniques for troubleshooting distillation columns.

550564 Radioisotope Techniques for Process Investigations

In all chemical plants and refineries there generally is not enough instrumentation in the right places in our processes. During the design phase, decisions were made between knowing more about the process, which means buying more instruments, and saving capital funds. When a problem appears with the operation of a vessel, there are always gaps in the knowledge of what is happening due to a lack of instruments in the places where knowledge is not routinely required. Process engineers and operations personnel need to know what is happening so they can take corrective action. When knowledge is missing, they often take actions that make matters worse.

For distillation and separation processes Chemical Engineers have historically relied upon plant process measurements such as flow rates, temperatures, and pressures and model or simulation results for data with which to solve troubleshooting or re-design problems. An array of on-line diagnostic services are available that can provide additional data offering real-time information on how pieces of process equipment are actually operating. The most common application is Gamma Scanning. This test is primarily applied, but not necessarily limited to, distillation or separation columns. Gamma scans provide a density profile of the internal process of operating distillation columns and other process vessels. The density profile can be used to diagnose the hydraulic operating conditions of mass transfer devices such as damage to internals, flooding, degree of entrainment or weeping, liquid levels on trays and distributors, liquid distribution through packed beds, etc. The presenter will show case studies where scanning revealed vital process information that helped solve an operating problem, or helped make a revamp/re-design successful. Additional derivatives from gamma scanning of detecting flow distribution patterns will be demonstrated. The ThruVision Scan technique is a specialized horizontal gamma scan used to generate a topographic profile of the internal cross-sectional density of process equipment. This profile is useful for the detailed study of liquid flow distribution through packed columns as well as having applications beyond distillation columns. A hybrid application involving tracers and scanning on fixed bed catalytic reactors will also be discussed.

AUTHOR INDEX